P9-CCF-136

Material Matters

Mixtures, Compounds & Solutions

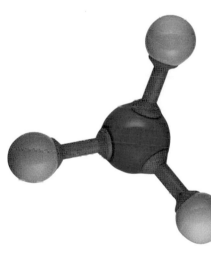

Carol Baldwin

Chicago, Illinois

For information, address the publishers:
Raintree, 100 N. LaSalle, Suite 1200, Chicago, IL 60602

Printed and bound in China
10 09 08 07
10 9 8 7 6 5 4 3 2

Library of Congress Cataloging-in-Publication Data

Cataloging-in-publication data is available at the Library of Congress

Baldwin, Carol, 1943-
   Mixtures, compounds & solutions / Carol Baldwin.
       p. cm. -- (Material matters)
   Includes bibliographical references and index.
   ISBN 1-4109-1677-4 (library binding-hardcover) -- ISBN 1-4109-1684-7 (pbk.)
   ISBN 978-1-4109-1677-8 (library binding-hardcover) -- ISBN 978-1-4109-1684-6 (pbk.)
   1.   Chemical elements--Juvenile literature. 2.   Chemicals--Juvenile literature. [1. Chemical elements. 2. Chemicals. 3. Mixtures.] I. Title: Mixtures, compounds, and solutions. II. Title. III. Series: Baldwin, Carol, 1943-   Material matters.
   QD466.B232 2005
   546--dc22

This leveled text is a version of Freestyle: Material Matters: Mixtures, Compounds, and Solutions.

Acknowledgments
Page 7, Tudor Photography; 20, Tudor Photography; 31, Tudor Photography; 41, Tudor Photography; 21, Art Directors & Trip; 42–43, Art Directors & Trip/ H Rogers; 30, Art Directors & Trip/M Walker; 6–7, Corbis; 8–9, Corbis; 10, Corbis; 11, Corbis; 12, Corbis; 14, Corbis; 22–23, Corbis; 24, Corbis; 34, Corbis; 36–37, Corbis; 38–39, Corbis; 24–25, Corbis/Bettmann; 34–35, Corbis/C Cohen; 20–21, Corbis/D Pebbles; 29, Corbis/G Lepp; 44, Corbis/G Lepp; 18, Corbis/J L Pelaez; 28–29, Corbis/ K Fleming; 43, Corbis/L Bergman; 30–31, Corbis/L Lefkowitz; 19, Corbis/M Gerber; 33, Corbis/O Franken; 25, Corbis/P Souders; 40–41, Corbis/S Agliolo; 15, Corbis/Vanni Archive; 9, Corbis/W White; 10–11, Empics; 6, FLPA; 5 bottom, FLPA/Foto Natura Catalogue; 35, FLPA/Foto Natura Catalogue; 4, FLPA/Maurice Nimmo; 32–33, FLPA/Minden Pictures; 17, FLPA/R Brooks; 22, FLPA/Robin Chittenden; 19, FLPA/W Meinderts; 27, FLPA/W Wisniewski; 4–5, Geophotos/T Waltham; 8, Geophotos/T Waltham; 14–15, Geophotos/ T Waltham; 23, Geophotos/T Waltham; 5 top, Photodisc; 16–17, Photodisc/; 36, Photodisc; 40, Photodisc; 45, Photodisc; 37, Science Photo Library; 38, Science Photo Library; 16, Science Photo Library/ Adam Hart Davis; 12–13, Science Photo Library/CNRI; 39, Science Photo Library/Earth Satellite Corporation; 42, Science Photo Library/Mark Thomas; 5 middle, Science Photo Library/ Prof S Cinti/CNRI; 26–27, Science Photo Library/Prof S Cinti/CNRI; 28, Trevor Clifford

Cover photograph reproduced with permission of Topham Picturepoint

Every effort has been made to contact copyright holders of any material reproduced in this book. Any omissions will be rectified in subsequent printings if notice is given to the publishers.

Contents

Any words appearing in the text in bold, **like this**, are explained in the Glossary. You can also look for some of them in the Word bank at the bottom of each page.

Climbing High

Granite
This is a piece of red granite. You can see the tiny bits or particles that make up the rock. Dark bits are mixed together with light bits.

Half Dome is a huge dome of granite rock. It is in Yosemite National Park, California. Half Dome is 4,752 feet (1,462 meters) high. Many people visit the Yosemite Valley each year and see the rock. But how many people think about what the rock is made of?

All rocks are **mixtures** of **minerals**. Most granite is made up of tiny bits of the minerals quartz, mica, and feldspar.

Granite from Half Dome is different than granite from other places. This is because granite is a mixture.

Word bank   mineral nonliving solid material

Compounds and mixtures

The minerals in rocks are chemical **compounds**. This means they are made from different types of **atoms**, or tiny **particles** joined together. The atoms in mixtures come together, too, but they do not bond. Most rocks are mixtures of compounds.

Solutions

Pure water is a compound. But most water has other materials mixed with it. Then water is a special type of mixture. It is a **solution**.

Find out later ...

... what is made when the Space Shuttle lifts off.

... what compound is found in your stomach.

... how can fish suffocate in water.

atom tiny particle that makes up all matter

Describing Matter

Five kinds of matter
The Ancient Chinese thought that there were five kinds of matter. These were fire, wood, metal, earth, and water.

Matter is everywhere. Anything that takes up space and has **mass** is matter. The more mass something has, the heavier it is. Matter can either be a **pure substance** or a **mixture**.

A pure substance is the same throughout. Water and other **compounds** are pure substances. Mixtures do not have the same features throughout. Granite is a mixture. Some parts of granite have more dark **particles** than other parts.

Elements

Compounds and mixtures are made up of **elements**. Elements are made of only one type of **atom**. They are the purest type of matter. Carbon is an element.

States of matter

Matter can take different forms. Most matter can be solid, liquid, or gas. These forms are called the **states of matter**.

Is this glass only two-thirds full? No! This glass is full. Part of the glass that does not contain orange juice contains air. Air is a mixture of invisible gases.

An iceberg is water in its solid state. Solids keep the same shape.

Orange juice is a liquid. It takes the shape of the glass.

Physical properties

Everything around us has features that help us to see what it is. Color, brittleness, and smoothness are some **physical properties**. You can usually tell these by looking or measuring.

Copper is an **element**. A piece of copper wire is shiny-orange and will bend. Chalk is a **compound**. A piece of chalk will break easily because it is **brittle**. It is white and dull. Parts of a **mixture**, such as sand and pebbles, have different colors, shapes, and sizes. So you can usually tell these by looking or measuring.

Glowing rocks
Some **minerals** glow when **ultraviolet (UV) light** shines on them. This is a physical property.

This piece of uranium oxide glows green in UV light.

Word bank   brittle firm to touch but easy to break

Physical changes

If you break a piece of chalk, its size and shape changes. Some of its physical properties have changed. But it is still chalk. A change in size or shape is a **physical change**.

A change in **state of matter** is also a physical change. When water **boils** or **freezes**, it is still water. When ice **melts** and changes from a solid to a liquid, it is still water. These physical changes do not change what makes up water.

Very, very hot! Diamonds have a very high **melting point**. The temperature must reach 6,440 °F (3,550 °C). Then a diamond will melt. This is two-thirds as hot as the surface of the Sun!

When a volcano **erupts**, hot, liquid rock pours out. This is called lava. When lava cools, it changes back to solid rock. This change of state is a physical change.

physical property feature that can be seen or measured without changing what a material is made of

Chemical properties

When a material is with another substance it may **react**. A **chemical property** tells us what will happen. Iron reacts with oxygen in damp air to form **rust**. This is a chemical property of iron.

A chemical property of fuels, such as gas and coal, is that they burn easily. Some **metals**, like calcium and zinc, react with **acids**. This is a chemical property of calcium and zinc.

Fall colors

In fall, leaves on some trees change color. This is because the chemical that makes them green changes. A color change is one sign of a chemical change.

chemical property feature of something that tells us how it will react

Chemical changes

A chemical change takes place when iron reacts with oxygen. A new material is formed. This is called iron **oxide,** or rust. This change is called a **chemical reaction**. The rust is a new material with new **properties**.

Elements, compounds, and **mixtures** all take part in chemical reactions.

Gasoline burns in a race car engine. This is a chemical property of gas. It gives off **energy** to make the engine run.

Crazy colors

To get color changes like this you have to use hair coloring kits. These kits contain three special chemicals. These chemicals react with the hair and change the hair's natural color.

Building Blocks of Matter

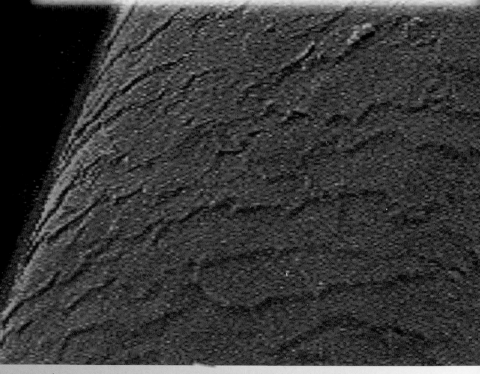

The Ancient Greeks decided that you could not keep on dividing **matter** forever. At some point you would reach the smallest piece. They named this smallest piece an **atom**. It means "cannot be divided."

In the early 1800s, John Dalton came up with some ideas about matter:

- all **elements** are made from atoms
- atoms cannot be divided or destroyed
- atoms of the same element are all the same
- atoms of different elements are different
- atoms of two or more elements can join to form **compounds**.

Atomic power
Today we can split the atom into even smaller particles. This releases huge amounts of **energy**. It is called nuclear energy. We can see this when an atom bomb explodes.

Word bank   nucleus center of an atom, made of protons and neutrons

Smaller particles in an atom

All matter is made of atoms. An atom has a center part. This is called a **nucleus**. The nucleus is made of two kinds of **particles**. **Protons** are particles with a positive charge. **Neutrons** are particles that have no charge.

Electrons are a third kind of particle in an atom. They have a negative charge. In an atom, the number of electrons is the same as the number of protons. The charges balance. So, overall, the atom has no charge.

Where are the electrons?

Electrons are arranged around the nucleus. They are in different energy levels or shells. Each energy level can only hold a certain number of electrons before it becomes full.

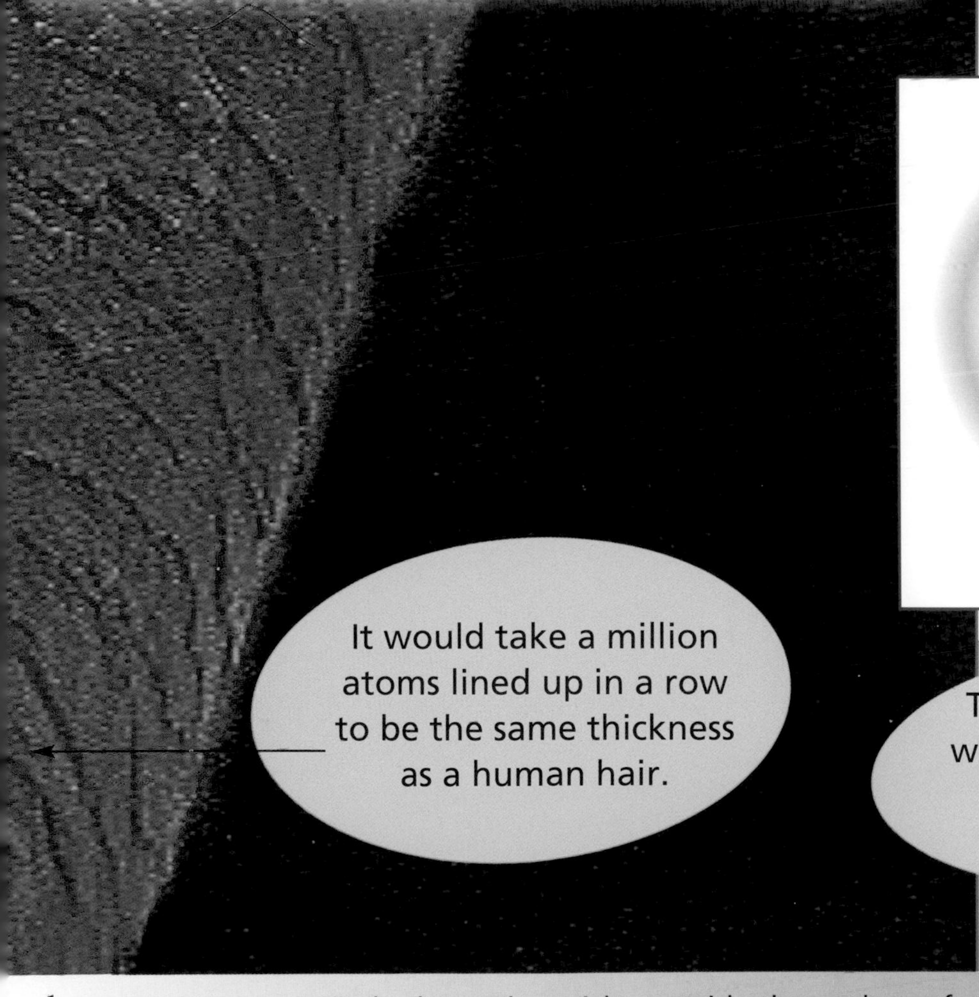

nucleus

electron levels

It would take a million atoms lined up in a row to be the same thickness as a human hair.

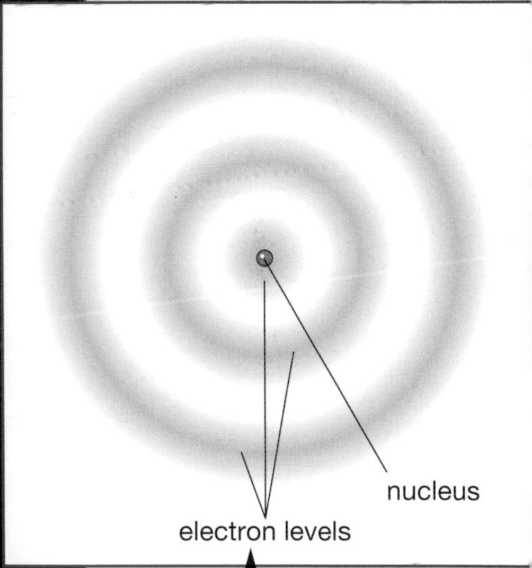

This diagram shows where scientists think electrons are in an atom.

electrons tiny, negatively charged particles outside the nucleus of an atom

Identifying elements

An **element** is made from only one type of **atom**. There are three main groups of elements.

Metals

Most elements, such as iron, are **metals**. Nearly all metals are solids. Only mercury is a liquid at **room temperature**.

All metals can look shiny. Heat and electricity move well through metals. Many metals can be stretched into wires. They can be hammered or shaped without breaking.

Metal ores

Metals are often found in rocks, called **ores**. Hematite is an ore. It is a compound of iron and oxygen. We mine hematite to get iron.

This **geyser** is in Yellowstone National Park. It has sulfur around its edge. Sulfur is a yellow, nonmetal element.

nonmetal element without the properties of metals

Nonmetals

Most **nonmetals** are gases, such as nitrogen. Some are solids, such as carbon. Diamonds are a form of carbon. Bromine is the only nonmetal that is liquid at room temperature. Nonmetals are not shiny. Heat and electricity do not pass through them easily. Solid nonmetals usually break if they are hammered. They are **brittle**.

Metalloids

A few elements have **properties** of both metals and nonmetals. They are called **metalloids**.

Symbols of elements

Chemical symbols are a short way of writing the names of elements. Symbols have one or two letters. The first is always a capital letter.

This is lead. Its chemical symbol is Pb. This comes from its **Latin** name, *plumbum*.

Compounds

Compounds are made of **atoms** of more than one **element**. Water is a compound. It is made of hydrogen and oxygen atoms. The atoms in a compound are joined tightly together.

Combining elements

Elements join together to make compounds only if there is a **chemical reaction**. When small pieces of iron are mixed with sulfur powder, you get a **mixture**. A chemical reaction happens only if this mixture is heated. Then a compound called iron **sulfide** forms.

How many compounds?

There are over one hundred elements. These combine in different ways. They make up over 21 million compounds.

This is a model of the compound ammonia. The white balls are hydrogen atoms, and the blue ball is a nitrogen atom.

compound   material made of two or more different types of atoms joined together

Properties of compounds

Compounds do not have the **properties** of the elements they contain.

Sugar is a white solid with a sweet taste. Sugar is a compound of hydrogen, oxygen, and carbon. Carbon is a black solid. Hydrogen and oxygen are both gases with no color. Hydrogen, oxygen, and carbon look very different from sugar.

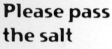

The Space Shuttle used aluminum powder as a fuel. Aluminum joins up with oxygen. The compound aluminum oxide is formed.

Please pass the salt

Sodium is a bright, silvery **metal**. It explodes when it **reacts** with water. Chlorine is a **poisonous** gas. But they react together to form white **crystals** of sodium chloride, or salt.

Carbon and oxygen

Carbon usually burns in air that has plenty of oxygen. Then carbon dioxide, CO_2, is formed. Carbon also burns when there is not enough oxygen. Then carbon monoxide, CO, is made.

A **compound** is always made up of the same **elements**. Also, the **atoms** always join up in the same way. If they join in a different way, a different compound forms.

Water has two hydrogen atoms to every oxygen atom. But hydrogen and oxygen can join in different ways. Two atoms of hydrogen can join with two atoms of oxygen. This forms a compound called hydrogen peroxide. Water and hydrogen peroxide are very different. Hydrogen peroxide is used to kill **bacteria** on cuts and to **bleach** hair.

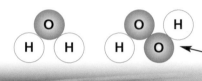

This diagram shows how atoms are joined in **molecules** of water, H_2O (left) and hydrogen peroxide, H_2O_2 (right).

Carbon monoxide is made in engines when there is not enough air.

molecule two or more atoms held together by chemical bonds

Breaking elements apart

You can break compounds apart. But you need a **chemical reaction** to do this. **Energy** is often needed to make a reaction happen. This could be heat, light, or electricity. If you heat sugar until it is very hot, it causes a chemical change. It breaks the sugar apart. If you heat sugar for long enough, only black carbon will be left. The hydrogen and oxygen join to form **water vapor**. This goes off into the air.

Living things, like mushrooms and molds, are decomposers.

Decomposers

There are lots of compounds in dead plants and animals. These will break down into elements and simpler compounds. Living things called decomposers are able to do this.

Hydrogen peroxide can make brown hair blonde.

Chemical bonds

Atoms are tightly joined together when they make a **compound**. **Chemical bonds** form between them. But not all chemical bonds are the same. Compounds belong to two different groups. These are based on their chemical bonds.

Covalent bonds

In some compounds, the atoms share **electrons** in a bond. These are called **covalent compounds**. Water is a covalent compound.

Harmful chemicals

Covalent compounds can be harmful. They turn into a gas easily. Many also burn easily. Covalent compounds are often used in nail polish remover.

Uses: As a fuel for burners and as a household solvent. Harmful by inhalation, in contact with skin and if swallowed. Harmful: possible risk of irreversible effects through inhalation, in contact with skin and if swallowed. Keep locked up and out of the reach of children. Keep container tightly closed. Keep away from sources of ignition - No smoking. Wear suitable protective clothing and gloves. In case of accident or if you feel unwell seek medical advice immediately (show the label where possible). Contains Methanol.

Products that are dangerous have warnings on their labels.

chemical bond **strong attraction between two atoms**

Ions and ionic bonds

Metals join with **nonmetals**. The metal gives one or more electrons to the nonmetal. Sodium is a metal. Chlorine is a nonmetal. When they join, sodium chloride, or salt, is formed. Each sodium atom gives an electron to a chlorine atom.

When sodium chloride forms, the sodium atom loses an electron. It has a positive charge. It is called a sodium **ion**. The chlorine atom takes an electron. It has a negative charge and is called a chloride ion.

Crystals
These are **crystals** of salt, or sodium chloride. Each crystal contains sodium and chloride ions.

Both the water and sand in this picture are covalent compounds.

ion atom or group of atoms with an electric charge

Nitrogen oxides

There are a few different nitrogen **oxides**. They are all gases. Nitrogen oxide, NO, comes from car exhausts. Dentists use dinitrogen oxide, N_2O, to relax patients. It is also called "laughing gas."

Chemical formula

All **elements** have a **chemical symbol**. These symbols are used for writing down a **formula**. A formula shows how many **atoms** of each element are joined together. The formula for oxygen is O_2. The O stands for oxygen. The small 2 tells us that two oxygen atoms are joined together. The formula for carbon dioxide is CO_2. The C stands for carbon. This shows that every carbon atom is joined to two atoms of oxygen.

Nitrogen dioxide, NO_2, is a **pollutant** in the air. It causes a brown haze. This is sometimes seen over large cities.

formula symbols and numbers to show how elements are joined

Naming compounds

There are a few rules for naming **compounds** that contain a **metal**. Here they are:

- put the metal's name first

- if the compound contains one nonmetal, change the end of the nonmetal's name to *ide*. For example, sodium (metal) and chlorine (nonmetal) **react** to form sodium chloride.

- if oxygen and another nonmetal are in the compound, change the ending of the nonmetal to *ate*. Potassium **carbonate** contains potassium, carbon, and oxygen atoms.

No metal at all
Some compounds are made of two nonmetals. They do not contain any metal. Then the second nonmetal ends in *ide*. For example, carbon monoxide and carbon dioxide are nonmetal compounds.

Carbon dioxide is in these soft drink bubbles. "Di-" means two. So, carbon dioxide has two oxygen atoms.

This limestone is calcium carbonate.

Important Compounds

Water in foods
Most foods contain water.

Food	Percent water
Celery	94
Tomatoes	93
Spinach, raw	92
Strawberries	90
Apples	85
Bananas	76
Eggs, uncooked	74
Macaroni, cooked	72
Chicken, grilled	71
Beef, raw minced	54
Ham, cooked	54
Bread, whole wheat	35
Honey	15

Water is the most important **compound** on Earth. Without water, all living things would die.

Water cycle
Water on Earth is always moving. The Sun heats water in lakes and oceans. It changes to **water vapor**. The water vapor rises into the air. Then it cools and forms clouds. Inside the clouds, tiny drops of water join to form bigger drops. These fall as rain, snow, or **hail**. Water runs off the land into rivers. This flows back into lakes and oceans. This is the **water cycle**.

Fresh fruit and vegetables are mainly water.

Word bank water cycle cycle of water from liquid to gas on Earth

Ores

Rocks often contain useful **metals**. These are called **ores**. Most ores are compounds of a metal and one or two **nonmetals**.

Oxides, sulfides, and carbonates

An ore of a metal joined with oxygen is called an **oxide**. Copper oxide is an ore. An ore of a metal joined to sulfur is called a **sulfide**. Lead sulfide is an ore. A **carbonate** ore is a metal combined with carbon and oxygen. Magnesium carbonate is an ore.

Getting metal from ore

The **mineral** coke is a form of carbon. When coke is heated, it takes the oxygen from the metal oxide. The metal is left over.

A blast furnace changes iron oxide to iron. Carbon dioxide is produced.

These cave paintings at Lascaux, France, were made using an iron oxide, called ocher.

ore metal that is combined with other elements

25

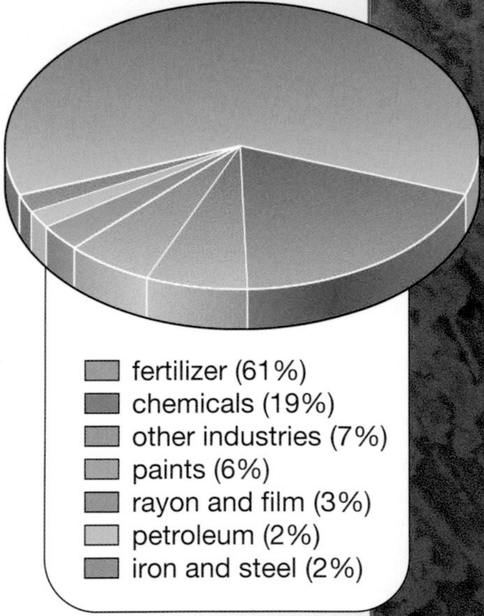

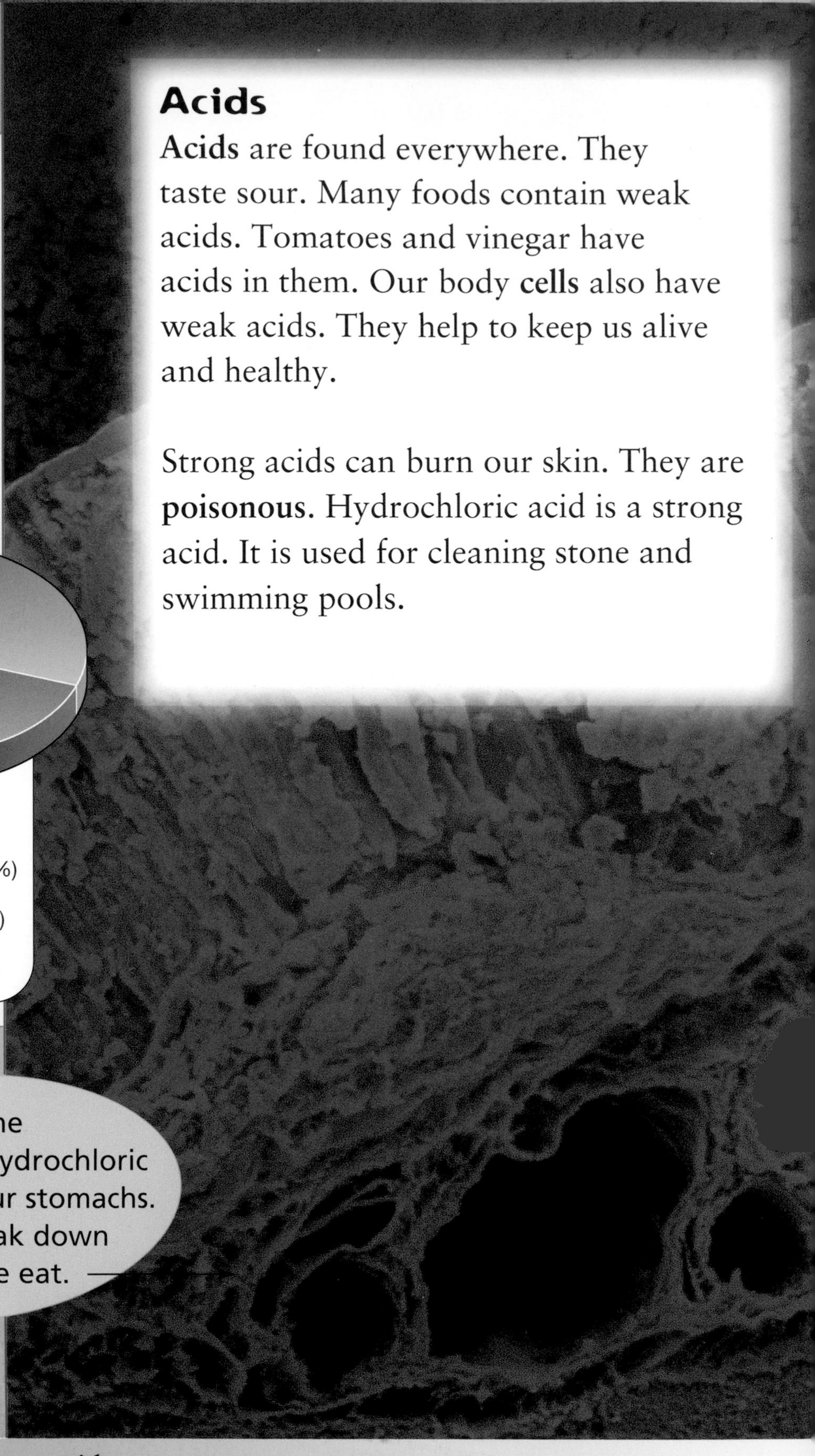

Acids

Acids are found everywhere. They taste sour. Many foods contain weak acids. Tomatoes and vinegar have acids in them. Our body **cells** also have weak acids. They help to keep us alive and healthy.

Strong acids can burn our skin. They are **poisonous**. Hydrochloric acid is a strong acid. It is used for cleaning stone and swimming pools.

Uses of sulfuric acid

Sulfuric acid is a strong acid. It has more uses than any other acid. This pie chart shows some of them.

fertilizer (61%)
chemicals (19%)
other industries (7%)
paints (6%)
rayon and film (3%)
petroleum (2%)
iron and steel (2%)

This is the stomach lining. Hydrochloric acid is found in our stomachs. It helps to break down the food we eat.

 Word bank acid compound that has a sour taste and can burn you

Bases

Bases are very common, too. They have a bitter taste and feel slippery. Strong bases are poisonous. They can burn you badly. Sodium hydroxide is also called caustic soda. It is strong enough to **dissolve** bones. Strong bases are good for cleaning drains and ovens. But you have to be careful with them.

Calcium hydroxide is a weak base. It is used in medicines called **antacids**. Antacids help soothe upset stomachs. They **react** with stomach acid.

Desert soils
Desert soil contains a lot of bases. In deserts, there is little rain to wash the bases away. Plants, like these cacti, can grow in the dry, **basic** desert soils. Many other plants cannot.

base compound that feels slippery and can burn you

Salts

Salts are common **compounds**. They form when an **acid reacts** with a **base**. Sodium chloride is the best-known salt. It has thousands of uses. Only a small amount is used on food. Meat packers **preserve** meat with salt. Salt is used to make glass, treat leather, and to make other chemicals.

Other salts are important, too. Ammonium chloride is used in batteries. Silver bromide is used in making film for cameras.

Epsom salts

Epsom salts are used for healing some skin rashes. They contain magnesium sulfate.

EPSOM SALTS B.P.
(Magnesium Sulphate B.P.)

Potassium nitrate is used to make explosives.

preserve stop from going bad

Compounds in living things

Nitrogen, oxygen, hydrogen, and carbon are found in **proteins**. Much of our body is made of protein. We need protein for growth. Meat, fish, and dairy products are sources of protein.

Carbohydrates are made up of carbon, hydrogen, and oxygen. Starch and sugar are carbohydrates. We need these for **energy**. Fruit, vegetables, and pasta contain sugar and starch.

Oils and fats are made up of carbon, hydrogen, and oxygen. We need them to keep our **cells** working. But too many can be harmful.

Cellulose

Cellulose is found in plants. It is a kind of carbohydrate. Humans cannot break down cellulose. But some animals can break it down into sugar. Cattle and termites, like those above, can do this.

protein   carbon compound that contains nitrogen, used for growth and repair in the body

Mixtures

Fabric mixtures

Many fabrics are mixtures. Your shirt label tells you what the mixture is. It may say the fabric is cotton and polyester. The shirt fabric has properties of cotton. But it will have some of polyester, too.

Most things around us are **mixtures**. Ocean water and fruit salad are mixtures. Mixtures are made up of several **pure substances**. But they are not held together by **chemical bonds**. They are separate. The pure substances in a mixture keep their own **properties**.

There are two types of mixtures. Ocean water is one type. You cannot see the different parts. The other type of mixture has larger parts that you can see. Rocks and fruit salads are like this.

A mixture can contain **elements**, **compounds**, or both.

You can see this fabric is a mixture of lycra and viscose fibers.

element	mixture of elements	compound	mixture of compounds

chemical bond strong attraction between two atoms

Separating mixtures

You can pick out the different parts in a mixture, if you can see them. Some mixtures contain solids of different sizes like rocks, pebbles, and sand. You can separate them by pouring the mixture through sieves with smaller and smaller holes.

Filtering can separate a sand and water mixture. The water passes through the filter paper, and the sand is trapped.

Brushing with a mixture

Toothpaste is a mixture. The materials are evenly spread throughout. But the materials in the toothpaste and their amounts vary. That is how different toothpastes can have different colors or flavors.

A mixture of sand, gravel, and other materials is mined at this quarry.

Solutions

Sugar seems to disappear when it is mixed in water. The pieces of sugar separate into sugar **molecules**. They are too small to be seen. We say the sugar **dissolves** in the water. This mixture is called a **solution**.

In a solution of sugar and water, the sugar molecules are spread evenly through the water. All solutions are like this.

Ocean water

Ocean water contains a lot of dissolved solids. The pie chart below shows them.

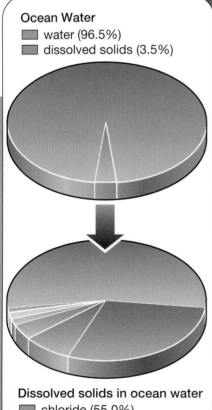

Ocean Water
- water (96.5%)
- dissolved solids (3.5%)

Dissolved solids in ocean water
- chloride (55.0%)
- sodium (30.6%)
- sulfate (7.7%)
- magnesium (3.7%)
- calcium (1.2%)
- potassium (1.1%)
- other (0.7%)

This shark is swimming in a solution. Ocean water contains many dissolved materials.

Word bank   dilute (solution) containing a small amount of material

Describing solutions

Solutions can be weak or strong. A weak solution has a small amount of material dissolved in it. This is called a **dilute** solution. A strong solution has a large amount of material dissolved in it. This is a **concentrated** solution.

As you add more material to a solution, it gets more concentrated. Finally, no more material will dissolve. The solution is **saturated**.

Stopping foods from going bad

Foods, like these hams, are **preserved** with salt. This removes water from the meat. Drier foods last much longer than fresh foods.

Dissolving solids in liquids

A **solution** is usually made from a solid **dissolved** in a liquid. The solid can dissolve at different speeds. It depends on how big the pieces of solid are. The smaller the pieces, the faster they dissolve. Stirring makes a solid dissolve faster, too.

The temperature of the liquid affects how much solid will dissolve. Hot water will dissolve more sugar than cold water.

Adding salt to ice cream
Salt is added to the ice in an ice cream maker. This makes the water **freeze** at a lower temperature.

freeze change from a liquid into a solid

Dissolving gases in liquids

Gases, like ammonia and oxygen, dissolve in water. They form a solution. Ammonia water is used for cleaning. Oxygen dissolves in lakes and oceans. Fish breathe the dissolved oxygen in the water.

Unlike a solid, less gas will dissolve if it is warm. Soft drinks have carbon dioxide gas dissolved in them. A glass of soft drink soon goes flat in a warm room.

Hot fish
Fish must have oxygen dissolved in their water. If the water becomes too warm, it loses oxygen. The fish might die.

You can see the **minerals** around the edges of this hot spring. The hot water is **saturated** with minerals. When it cools, the minerals come out of the water.

Dissolving liquids in liquids

Many materials **dissolve** in water. These include other liquids. Alcohol is a liquid that will do this. Antifreeze is a **solution** of water and alcohol. People put it in their car radiators. It stops the water in the radiator from **freezing**.

Not all liquids dissolve in water. For example, if you mix oil and water, oil floats on top of the water.

Making citrus oils

There is oil in the peels of lemons and oranges. The juice and oil is squeezed out of the fruit. The oil rises to the top of the juice. Then it can be separated.

You should not use water to put out burning oils or gases. They do not dissolve in water. Spraying water on them spreads the fire.

dissolve mix completely and evenly

Separating solutions

It is not hard to separate the parts of a solution. You can separate a solid dissolved in a liquid just by leaving it in the air. The liquid will **evaporate** and leave the solid behind.

When a gas is dissolved in a liquid, you can heat the liquid. The gas escapes into the air. The liquid is left behind.

It is easy to separate solutions of liquids, too. The **mixture** is heated slowly. Each liquid changes to a gas at a different temperature. As each liquid boils off, its gas is collected. The different parts of air are separated like this.

Air is cooled until it becomes a liquid. Then it is warmed up again very slowly. Nitrogen **boils** at a lower temperature than oxygen, so it turns into a gas first. This leaves oxygen behind. The two gases are collected separately.

Black ink is made up of a mixture of colors. You can separate the colors like this. This method is called **chromatography**.

Separating colors

In this photograph, ink from a marker was placed near the bottom of the **filter** paper. The ink was dried. The paper's bottom edge was placed so it just touched the liquid. The liquid moved up the paper. This separated the colors in the ink.

evaporate change from a liquid to a gas

More Mixtures

This blood was spun at very high speeds. The suspended particles are pulled to the bottom of the tube.

When you put pond water into a jar, it looks muddy. What happens if you let the **mixture** stand? Soil **particles** settle to the bottom of the jar. When soil particles do this, they are called **sediment**.

Particles in a suspension

You can easily separate sediment from the liquid above it. You just pour it off carefully. You do this at home when you pour out the liquid from a can of tuna fish.

A mixture of soil and water is a **suspension**. A suspension is a cloudy mixture of materials. The materials always separate when left to stand.

Fast water, like this, wears away bits of soil and rock from the land. These materials are suspended in the water.

Blood
Blood is a **solution**. Blood contains **dissolved** solids and gases. This is the yellow part.

Blood is also a suspension. Red blood **cells** and white blood cells are some of the solids suspended in the blood.

sediment solids not dissolved in a liquid. They fall to the bottom of the liquid.

Separating suspensions

If you leave a suspension to stand, the particles will settle to the bottom. But large particles settle out more quickly than small particles. Sand and clay form a suspension in water. The sand settles to the bottom in minutes. The tiny clay particles will take hours to settle to the bottom.

You can separate a suspension quickly. You pour it through a **filter**. Filters are often made from paper. They have tiny holes. Particles that are larger than the holes cannot pass through. Most of our drinking water is filtered.

River deltas

A river carries lots of sediment. When the river empties into an ocean, it slows down. Then the materials settle to the bottom. Some rivers drop so much material that the sea cannot wash it away.

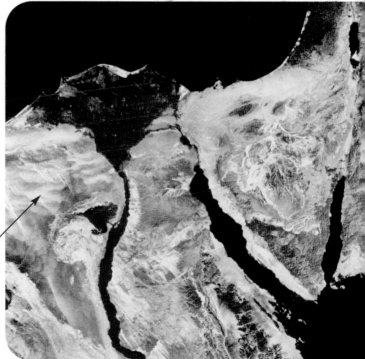

The red area is sediment. It forms new land called a delta. This is shaped like a fan, and reaches into the ocean.

suspension mixture that contains a liquid in which particles settle over time

Colloids

Whipped cream, fog, and smoke are all **colloids**. A colloid is a **mixture**. Its **particles** do not separate when it is left to stand, like they do in a **suspension**. The particles in a colloid stay mixed. This is because they are always bumping into one another.

The particles in a suspension are quite large. The particles in a **solution** are tiny. The particles in a colloid are in between.

Cleaning with a colloid

A colloid is used to clean our water. Then it can be used again. The colloid looks like a fluffy gel. **Bacteria** and tiny particles stick to the gel and are trapped.

This is a sewage treatment plant. A colloid is used to trap bacteria.

colloid mixture that has a particle size between that of solutions and suspensions

Kinds of colloids

Whipped cream is a colloid of a gas in a liquid. It is air in cream. Fog and clouds are colloids of a liquid in a gas. These are water droplets in air. Smoke is a colloid of solids in a gas, which is air. **Gels**, like jelly, are colloids of solids in a liquid.

Colloids can scatter light. If a beam of light is shone through a colloid, you can see it. Car headlight beams can be seen in fog. The water droplets in air scatter the light.

Mixing milk
In fresh milk, the cream rises to the top of the bottle. Dairies break the cream into very small particles. The cream forms a colloid with the rest of the milk. Then it stays mixed. The milk is **homogenized**.

You can see this lighthouse beam. This is because of the tiny particles of dust and water in the air.

gel colloid of solids in a liquid

Alloys

Mouth metals

Dentists use an alloy to fill holes in teeth. It is called **amalgam**. It contains mercury. When it is first made, amalgam is a liquid. But it hardens quickly.

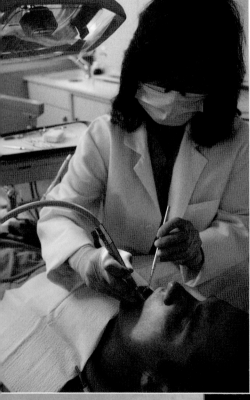

Alloys are **mixtures**. They contain a **metal** and one or more other **elements**. Aluminum alloys are used to make aircraft. They are strong and lightweight.

Bronze is an alloy of copper and tin. Bronze is used for things that may come into contact with salt water. The tin does not **react** with the chemicals in seawater. Brass is an alloy of copper and zinc. It is used in musical instruments.

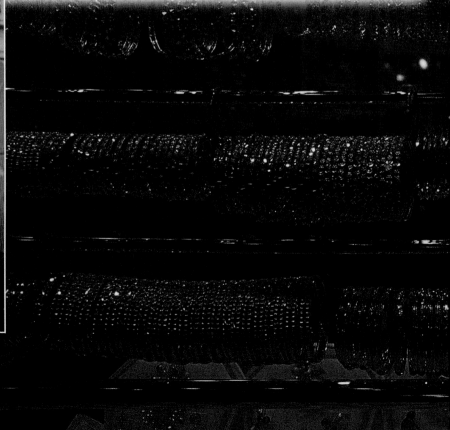

Word bank   alloy mixture of a metal and one or more other elements

Steel

There are many types of steel. They are all alloys of iron and other elements.

Different steels are made for different uses. Steel with manganese in it is very hard. It is used in railroad tracks and military tanks. Stainless steel contains nickel and chromium. It is strong and does not **corrode**. Doctors use stainless steel tools for surgery.

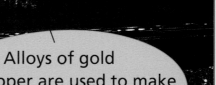

Alloys of gold and copper are used to make jewelry. The copper adds strength to the gold.

Mending bodies

Alloys are used to mend broken bones, as this X-ray shows. Stainless steel was used to make plates and screws. Now newer alloys are used instead. They are lighter and stronger than steel.

corrode damage by a reaction with chemicals

Find Out More

American Chemical Society
The American Chemical Society has members from all areas of chemistry and at all levels. It has a division especially for school students. This provides teacher and student resources, competitions, a magazine, a Web site, meetings, and help to get vacation jobs in chemistry.

Contact them at the following address:
American Chemical Society, 1155 Sixteenth Street, NW Washington, D.C., 20036

Books

Oxlade, Chris. *Chemicals in Action: Elements and Compounds*. Chicago: Heinemann Library, 2002.

Snedden, Robert. *Material World: Materials Technology*. Chicago: Heinemann Library, 2001.

Snedden, Robert. *Material World: Separating Materials*. Chicago: Heinemann Library, 2001.

World Wide Web

To find out more about compounds, mixtures, and solutions you can search the Internet. Use keywords like these:

- compounds +properties
- (name of a metal) +ore +mining
- mixtures +separating

You can find your own keywords by using words from this book. The search tips on the following page will help you find useful Web sites.

Search tips

There are billions of pages on the Internet. It can be difficult to find exactly what you are looking for. These tips will help you find useful Web sites more quickly:

- know what you want to find out about
- use simple keywords
- use two to six keywords in a search
- use only names of people, places, or things
- put double quote marks around words that go together, for example, "chemical formulas"

Where to search

Search engine
A search engine looks through millions of Web site pages. It lists all the sites that match the words in the search box. You will find the best matches are at the top of the list, on the first page.

Search directory
A person instead of a computer has sorted a search directory. You can search by keyword or subject and browse through the different sites. It is like looking through books on a library shelf.

Glossary

acid compound that has a sour taste and can burn you

alloy mixture of a metal and one or more other elements

amalgam alloy that contains mercury

antacid medicine that contains a weak base and is used to soothe upset stomachs

atom tiny particle that makes up everything

bacteria tiny living things, so small you need a microscope to see them

base compound that feels slippery and can burn you

bleach material that can remove color

boil rapid change of state from a liquid to a gas. This change takes place within the liquid and at its surface.

brittle firm to touch but easy to break

carbonate compound that contains carbon and oxygen

cells building blocks that make up all living things

chemical bond strong attraction between two atoms

chemical property feature of something that tells us how it will react

chemical reaction change that produces one or more new materials

chemical symbol short way of writing the name of an element

chromatography method used to separate colors in a mixture or a dye

colloid mixture that has a particle size between that of solutions and suspensions

compound material made of two or more different types of atoms joined together

concentrated (solution) containing a large amount of material

corrode damage by a reaction with chemicals

covalent compound compound formed when atoms join and share electrons

crystal solid that has particles arranged in a regular, repeating pattern

dilute (solution) containing a small amount of material

dissolve mix completely and evenly

electron tiny, negatively charged particle outside the nucleus of an atom

element material made from only one kind of atom

energy ability to cause change

erupt when a volcano shoots out lava violently

evaporate change from a liquid to a gas

filter strain by pouring through paper or cloth

formula symbols and numbers to show how elements are joined

freeze change from a liquid to a solid

gel colloid of solids in a liquid

geyser opening in Earth that shoots up water and steam from underground

homogenize break cream into tiny particles so that they do not separate from the milk

ion atom or group of atoms with an electric charge

Latin ancient language used by the Romans

mass amount of matter in an object

matter anything that takes up space and has mass

melt change from a solid to a liquid

melting point temperature at which a solid turns into a gas

metal material that is shiny and lets heat and electricity move through it easily

metalloid element with some properties of both metals and nonmetals

mineral nonliving solid material

mixture material made of elements or compounds not joined chemically

molecule two or more atoms held together by chemical bonds

neutron particle with no charge, found in the nucleus of an atom

nonmetal element without properties of metals

nucleus center of an atom, made of protons and neutrons

ore metal that is combined with other elements

oxide compound formed when oxygen joins up with another element

particle tiny bit

physical change change in how something looks, not in what makes it up

physical property feature that can be seen or measured without changing what a material is made of

poisonous harmful, as in a material that will harm you

pollutant harmful material in the air, water, or on the land

preserve stop from going bad

property feature of something

protein carbon compound that contains nitrogen, used for growth and repair in the body

proton positively charged particle in the nucleus of an atom

pure substance matter that is the same throughout

react take part in a chemical reaction and produce one or more new substances

room temperature about 68 °F (20 °C)

rust iron oxide, formed when iron reacts with oxygen in the air

salt compound formed when an acid reacts with a base. Sodium chloride is a common salt

saturated containing as much material as can be dissolved at that temperature

sediment solids not dissolved in a liquid. They fall to the bottom of the liquid.

solution mixture in which one material dissolves in another

state of matter whether something is solid, liquid, or gas

sulfide compound that contains sulfur

suspension mixture that contains a liquid in which particles settle after a time

ultraviolet (UV) light invisible light that is beyond violet in the spectrum and causes skin to burn

water cycle cycle of water from liquid to gas on Earth

water vapor water in a gas state

Index